Evolution
versus
Intelligent Design

Why all the Fuss?

The arguments for both sides

Peter Cook

Evolution
versus
Intelligent Design

Why all the Fuss?

The arguments for both sides

Peter Cook

First published in Australia in 2006 by
New Holland Publishers (Australia) Pty Ltd
Sydney • Auckland • London • Cape Town

www.newholland.com.au

14 Aquatic Drive Frenchs Forest NSW 2086 Australia
218 Lake Road Northcote Auckland New Zealand
86 Edgware Road London W2 2EA United Kingdom
80 McKenzie Street Cape Town 8001 South Africa

National Library of Australia Cataloguing-in-Publication Data:

Cook, Peter, 1963- .

    Evolution versus intelligent design : all the issues.

    ISBN 1 74110 471 8.

    1. Evolution. 2. Intelligent design (Teleology). 3.

    Religion and science. I. Title.

    213

Publisher: Martin Ford
Production controller: Linda Bottari
Project Editor: Michael McGrath
Designer: Barbara Cowen
Cover Design: Greg Lamont
Printer: McPherson's Publishing Group, Maryborough, Victoria

Contents

For Clementine and Florence

Introduction

WHY READ THIS BOOK?

If you have a school-aged child, or are simply interested in the debate about the teaching of evolution and intelligent design, your local library or the internet can provide you with a seemingly limitless supply of reading materials on the topic.

All of this literature, however, is written by someone who wants to convince you of their point of view. Be they a clergyman or a biologist, the author will set out arguments in favour of their own view, and present their opponents' arguments in ways often intended to make them look foolish. That is not the approach taken here.

This book aims to present both sides of the argument about evolution and intelligent design without bias, in order that you, the reader, can formulate your own view.

Both theories are explained in detail, the arguments for each viewpoint are presented as strongly as they can be, and where possible these arguments are put side by side for easy comparison.

The final chapter is titled 'Where to stand?' but it does not advocate a particular position on the debate. Rather, it simply explores the positions available for you, the reader, to take.

Why all the Fuss?

In the Biology Class

Until as recently as 2006, if you were a high school student in the United States, there was every chance that you were learning things in your biology class that were quite different from what students down the road or in the neighbouring district were learning.

Specifically, you might have been taught that evolution is just one explanation of the origin of life and that there are alternative theories, such as the theory of intelligent design (ID).

This difference in syllabus was not due to some startling breakthrough in a science laboratory. Rather, it was the result of a decision reached by the local school board. The members of this board are not likely to be practising scientists, but they *are* likely to have strong personal convictions about the value of Darwin's theory of evolution.

The debate over Darwin's theory of evolution has a long history, which has unfolded largely on American soil. The issue has now reared its head in Australia.

Let us begin, however, in Tennessee ...

HISTORY OF THE DEBATE

The Scopes Monkey Trial

Sixty-six years after Darwin published *On the Origin of Species* in 1859, John Scopes, a Tennessee high school biology teacher, was charged with illegally teaching the theory of evolution.

Teaching the theory had been banned in Tennessee that year and the infamous 'Monkey Trial' of 1925 attracted enormous publicity.

Scopes was convicted, although his conviction was later overturned. Tennessee's ban on teaching the theory of evolution was adopted by two other southern states and only repealed fifty years ago, in 1967.

The Rise of Science

At the time of the Monkey Trial, the theory of evolution was removed from the Tennessee school syllabus because it was deemed incompatible with the account of creation given in the Bible. Whether or not the idea of evolution was scientifically correct was not the issue.

Come the 1960s the Soviet Union launched the *Sputnik* satellite and the 'space race' was on. This motivated a crash program to bring American science education up to par, which included the use of a new evolution-based biology text book in nearly half the schools in the country.

Creation Science

This esteem of science affected proponents of Biblical creation as well. From the 1970s onwards, creationists argued that God's creation was something that could be scientifically studied by considering, for example, how the great flood of Noah's time has affected Earth's geology.

In Louisiana and elsewhere, the discipline of Creation Science won state legislative victories requiring that equal time and emphasis be given, in the classroom, to Bible-based accounts of creation as was given to evolution.

Balanced Treatment

These laws requiring 'balanced treatment' were proposed, as well as challenged, in Tennessee, Arkansas, Mississippi and elsewhere, until Louisiana creationists took their appeal for balanced treatment to the United States Supreme Court in 1987.

The Court ruled that Creation Science was an attempt 'to restructure the science curriculum to conform with a particular religious viewpoint.'

This, the Court concluded, was in violation of the First Amendment of the Constitution, which prohibits the use of government institutions to achieve a religious purpose.

As a result of this decision all existing 'balanced treatment' laws were deemed unconstitutional and thrown out.

THE INTELLIGENT DESIGN MOVEMENT

Origins of ID

In his 1991 book *Darwin on Trial*, Californian law professor Phillip Johnson used the name 'Intelligent Design' to describe an alternative to the theory of evolution. Johnson is usually credited with having founded the ID movement.

Most ID literature emerges from the Centre for Science and Culture (CSC), which is a subdivision of the Discovery Institute, a conservative Christian think tank based in Seattle. The CSC views ID as part of a larger program to overthrow 'materialist science' and replace it with 'theistic science'.

We shall consider what these ideas amount to in due course.

The Theory of ID

The theory of ID aims, firstly, to point out flaws in the theory of evolution. The second aim, pursued most notably in a number of books by Pennsylvanian professor of biochemistry, Michael Behe, and Kentucky philosopher, mathematician and seminary professor, William Dembski, both senior fellows at the CSC. They argue that some biological features, such as cells or bacteria, are of such complexity that they are best explained as the handiwork of an intelligent designer, rather than as products of evolution.

A notable feature of this viewpoint is that it is not based on religious assumptions: it does not assume from the outset that there is a god. Of course the viewpoint may have religious implications, but it aims to be fully scientific in its methods, simply studying biological life and attempting to explain it, as well as challenging other attempted explanations of it.

Assessing the Claims of ID

Now this scientific approach could be seen as nothing more than an attempt to get around the First Amendment, and have Christian ideas (the idea of a designer) reintroduced into the biology classroom.

Whatever its agenda, however, a viewpoint that makes scientific claims deserves to have those claims assessed on their scientific merits.

That is the purpose of this book; or rather this book aims to present the ideas at stake in the debate between evolution and ID, so that you, the reader, can make your own assessment of them.

Before turning to this, though, let us consider the reception ID has received in different sectors of society since its inception.

ID IN AMERICAN SCHOOLS

After 1991, school boards across the United States continued trying to ban or inhibit the teaching of evolution, but in some cases they also proposed ID as an alternative. In two cases this resulted in board members losing re-election bids.

In 1999, the Kansas Board of Education deleted any mention of evolution from the science curriculum, while in 2004, the School District Board in Dover, Pennsylvania, decreed that 'students will be made aware of gaps/problems in Darwin's theory and of other theories of evolution including intelligent design.' In both cases the board members advocating these views were not re-elected to office.

However, in 2004, the local school board in Blufton, Indiana announced that their school science curriculum would continue to require that teachers give a 'fair and balanced' presentation of 'appropriate theories' such as ID and evolution.

In 2005, the New York State Assembly introduced a bill calling for instruction in ID for all public-school students. The Kansas State Board of Education, too, has approved new state science standards, drafted by supporters of ID, which encourage school-teachers to challenge Darwinism.

Similar proposals have been passed in school districts in Ohio, Minnesota, and New Mexico.

Stickers on Textbooks

In Alabama, biology textbooks carry a warning which says that evolution is: 'a controversial theory some scientists present as a scientific explanation for the origin of living things. No one was present when life first appeared on Earth. Therefore, any statement about life's origins should be considered as theory, not fact.'

In 2003, the school board in Cobb County, Georgia followed suit with a sticker stating, 'This textbook contains material on evolution. Evolution is a theory, not a fact, regarding the origin of living things. This material should be approached with an open mind, studied carefully and critically considered.'

Parents in Georgia objected and, in 2005, a district judge ordered the removal of the stickers on the grounds that by denigrating evolution 'the school board appears to be endorsing a well-known alternative theory'—that is, intelligent design.

After the Cobb County case, the school district in Beebe, Arkansas removed stickers that had been on their textbooks for over a decade, which described evolution as 'controversial' and referred to an 'intelligent designer' as a possible explanation for the origin of life. Authorities in Alabama, however, did not follow suit, saying they didn't see how the ruling in Georgia applied to them.

The Latest Legal Finding

In late 2004, the School District Board in Dover, Pennsylvania mandated the teaching of ID in the biology curriculum from 2005 onwards.

Concerned parents took the matter to court and, in December 2005, Federal Court, Judge John E Jones, ruled that ID must meet the same fate Creation Science met in 1987.

ID, he ruled, 'is not science and is essentially religious in nature,' hence, like Creation Science, it violates the First Amendment. Experience, however, suggests that this will not be the end of the debate.

The President and the Vatican

In August 2005, US President George W. Bush answered a question about ID, saying, 'Both sides ought to be properly taught, so people can understand what the debate is about. Part of education is to expose people to different schools of thought. You're asking me whether or not people ought to be exposed to different ideas, and the answer is yes.'

In November 2005, Cardinal Paul Poupard of the Vatican responded that the Biblical account of creation and Darwin's theory of evolution were 'perfectly compatible' if the Bible were read correctly. Poupard explained that *Genesis* is not meant to be a scientific description. Rather, its message is simply that 'the universe didn't make itself, it had a creator.'

OPINIONS IN AMERICAN SOCIETY

A 2005 survey by Harris Polling found that most Americans (54%) do not believe that humans evolved from earlier species and 64 per cent believe that humans were created by God.

A poll by the Pew Research Centre seems to find a more even split between those who hold that 'living things have existed in their present form since the beginning of time' (48%), and those who believe that 'humans have evolved over time' (42%).

It is worth noting, however, that of those who hold that humans have evolved, over half hold that this evolution was guided by a 'supreme being' rather than by the Darwinian principle of natural selection.

ID IN AUSTRALIA

Religion or Biology?

Unlike the United States, Australia has no recognised institution or research centre devoted to developing and promoting ID. Nevertheless, the theory is developing a public profile, presumably as a result of US influence. In August 2005, the then Federal Minister for Education, Brendan Nelson, said that ID 'could have a place alongside evolution in our schools if parents wished.' He later clarified that ID should be discussed in religious studies rather than biology classes. Nevertheless, two months later, 70 000 Australian scientists and educators felt compelled to publicly condemn the teaching of ID in science classes.

Despite this, Pacific Hills Christian School in Dural, NSW is now teaching ID in science classes. Indeed, the chief executive of Christian Schools Australia, Stephen O'Doherty, said that ID was likely to be discussed in science classes in many Christian schools. ID is also being incorporated into Catholic school texts, albeit as a topic in religious studies. Australian Seventh Day Adventist Schools state that they too, 'will continue to teach intelligent design alongside the mandatory curriculum.'

ID is set to make further headway in Australia. Brendan Nelson made his comment the day after he met with a Christian group, called Campus Crusade for Christ. They gave him a copy of a DVD produced by the Discovery Institute in the United States, entitled *Unlocking the Mysteries of Life*. This DVD is a slick production featuring impressive computer graphics and an array of prominent scientists, who each explain why they no longer support Darwin's theory of evolution. This DVD is being distributed by Campus Crusade for Christ to every high school in Australia.

WHAT IS AT STAKE?

Teach the Controversy?

Having failed to secure a sanctioned place for ID in biology classrooms, proponents of the theory in the United States are advancing a campaign called 'Teach the Controversy'. Teachers are being urged to expose students to 'scientific arguments for and against Darwinian theory'.

Advocates of ID say that teaching students to 'critically analyse' evolution will help give them the skills to 'see both sides' of scientific issues. While this idea of honing student's critical skills seems fair-minded, to study arguments against evolution would be, precisely, to study ID: ID would have made it into the classroom after all.

Undermining Science

The American Association for the Advancement of Science argues that teaching ID as a criticism of evolution will not hone students' critical skills. Rather, they claim that it will 'undermine scientific credibility and the ability of young people to distinguish science from non-science.' Rhode Island biology professor and vocal opponent of ID, Kenneth Miller, argues that ID has the potential to drive people away from science: 'If classrooms are allowed to become theological battlegrounds, then schoolchildren will basically be told that science is hostile to new ideas and that scientists believe in a ludicrous theory that negates the very existence of God.'

Ironically, it seems that the aim of proponents of ID, in getting the theory taught in biology classrooms, is not simply to improve the science curriculum, but rather to make students more critical of science itself. This is because, on this view, science is seen as a form of 'materialism' that has taken over modern culture and robbed human life of its meaning, purpose and dignity (see page 65). Stephen Meyer, historian and director of the ID think tank at the Discovery Institute in Seattle, says, 'Our culture has been deeply influenced by materialist thought. We think it's deeply destructive, and we think it's false. And we mean to overturn it.'

Evolution is definitely seen by some as an example of this sort of destructive thought. Marc Looy, Vice President for Ministry Relations at the Kentucky creationist ministry Answers in Genesis, says, 'students in public schools are being taught that evolution is a fact, that they're just products of survival of the fittest. It creates a sense of purposelessness and hopelessness, which I think leads to things like pain, murder and suicide.'

Much, then, is at stake in the debate between evolution and ID. On the one hand ID is said to compromise science; on the other, evolution is said to rob life of its meaning and dignity. The chapters that follow consider each theory in turn, and what is at stake for each. On the basis of this, we then analyse the arguments put forward by these two viewpoints, so that you can work out where you stand on this issue.

What is Evolution?

The theory of ID is based, at least in part, on arguments that the theory of evolution is not adequate as an explanation of the origin and development of life. In order to understand the debate, let us begin by considering the theory of evolution. The theory is based on a few quite simple ideas. The critical question to keep in mind in this chapter is—can these simple ideas really account for the fantastic diversity and complexity of earthly life?

Charles Darwin

Charles Darwin was born in 1809, in Shrewsbury, England. The son of a doctor, Darwin initially followed in his father's footsteps, studying medicine in Edinburgh. However, Darwin spent more time studying the natural world than medicine, so his father moved him to Cambridge to study theology, in preparation for a career as a clergyman in the Church of England. Many Anglican parsons were naturalists, exploring the wonders of God's creation, and Darwin became an expert on beetles.

At this time, Darwin was enthusiastic about the idea of 'divine design' in nature, made popular by the English theologian William Paley.

That changed after 1831, when Darwin set sail on a five-year scientific expedition to South America aboard *HMS Beagle*.

Developing the Theory

The *Beagle* sailed around the world, crossing the Atlantic to the Americas and then home to England via the Pacific Ocean. In South America, Darwin found fossils of extinct animals that were similar to modern species.

On the Galapagos Islands in the Pacific, he noticed many variations among plants and animals of the same general type as those of South America. Darwin collected specimens everywhere he went, taking them home for further study.

On his return to England, Darwin continued his research and eventually formulated his theory of evolution in 1838. Darwin did not go public with his theory immediately. It was only when he got wind of a similar theory being proposed by fellow naturalist Alfred Wallace that, in 1859, Darwin published *On the Origin of Species by Means of Natural Selection, or The Preservation of Favoured Races in the Struggle for Life*.

DESCENT WITH MODIFICATION

The basic idea of evolution has two parts. The first is that plants and animals, indeed all life forms, inherit characteristics from their parents. Secondly, over time, and down through successive generations, these characteristics change.

In Darwin's time it was unclear how a parent managed to pass on characteristics to its offspring, but this inheritance is now understood as the result of the offspring inheriting the genes, or DNA, of its parents.

The genes that any individual inherits are the blueprint for the growth of that individual, and it is by modification to this genetic blueprint that a species, over time, evolves.

The Tree of Life

This idea of descent with modification suggests a particular story about how the diversity of life-forms on our planet originated. According to this story, in the distant past some simple organism came into being, and reproduced.

Once this pattern of reproduction was in motion, at some point some of the resultant offspring differed from their parents and siblings. And through further generations, more modification and variety occurred. This descent with modification resulted, ultimately, in the varied species of plants and animals that we see today.

A map of this development would look like a family tree. The first organism that long ago emerged from the primordial swamp would be at the base of the trunk. The successive modifications would eventually yield such branches as bacteria, fungi, plants and animals, and these branches would split again. The animal branch, for example, would branch further into fish, amphibians, primates, rodents and birds, and these in turn would branch into their own variety of species.

Mapping the Branches:
Shared Derived Characters

To reconstruct a branch of descent on this tree of life, biologists collect data about the characteristics of each organism they are interested in, noting such things as whether an animal, for example, has vertebrae, a bony skeleton, four limbs or an amniotic egg (an egg which contains fluid).

What the biologists are looking for is what they call 'shared derived character'. These are characteristics that a number of species share, but which are not apparent in other contemporary species or in earlier species.

For example, although they all possess a bony skeleton, amphibians, turtles, lizards, crocodiles, birds and mammals have four limbs, whereas ray-finned fishes do not. What this suggests is that at some time in the past the ancestor of these beasts with a bony skeleton gave birth to offspring with some modification that would eventually result in the development of four-limbed animals. At this point, then, there was a fork in the evolutionary road, with some of this ancestor's progeny heading down the four-limbed path, while others did not, eventually evolving into the ray-finned fishes.

Homologies and Analogies

Characteristics which are shared by a number of species due to their common ancestry are called 'homologies'. Having four limbs is a homology shared by birds, bats, mice and crocodiles, but not all shared characteristics are homologies. For example, bats and birds both have wings, but these do not seem to be something that they both inherited from a common ancestor with wings. A bat's wings are flaps of skin stretched between the bones of the fingers and forelimbs, whereas bird wings consist of feathers extending along the length of their forelimbs.

These structural differences suggest that bat and bird wings evolved independently. They only appear similar because they evolved to serve the same function: flying. Hence they are analogies, rather than homologies.

Dating the Tree of Life

Scientists estimate that there has been some form of life on earth for well over three billion years, but the evolution of multi-cellular life is, in terms of this scale, only very recent. If the length of each branch reflects the amount of time that branch existed before forking, the tree of life has a very long trunk before it reaches the first plant and animal branches. Scientists make claims about the age of branches on the basis of three sorts of evidence:

1 **radiometric dating**, which measures the decay of radioactivity in rocks and other matter
2 **stratigraphy**, which analyses strata in the earth's crust to establish the sequence in which things happened in the earth's history
3 **molecular clocks**, which allow scientists to measure the genetic divergence between organisms, and thereby estimate their age.

MECHANISMS OF MODIFICATION

Darwin's idea of descent or inheritance is easy enough to grasp. His idea of modification is not so obvious. Why is it that a parent can give birth to offspring which differ from their parents, and from each other? Moreover, how is it that such a chance inheritance comes to be perpetuated in further generations? Shouldn't chance modifications simply die out, rather than flourish?

Darwin names four mechanisms of modification. The first two, mutation and migration, are the means by which genetic variation occurs. The second two, genetic drift and natural selection, are the ways in which some of this variety flourish and others do not.

Modifications which can be Inherited

For a modification to be inheritable there must be a change in the genetic makeup of some portion of a population of plants or animals. Imagine some beetles whose food supply is restricted due to drought. These beetles may be smaller than the previous generation. Is this an example of evolution? No, it is not. The beetles are smaller because of a lack of food, which is a feature of their environment, rather than a result of their genetic makeup. When the drought is over the next generation of beetles will return to their normal sizeSuppose now that a few of the beetles were brown rather than green, and that a generation or two later, there were more brown beetles than green. In this case there has been a change in the genetic makeup of the population, so this is an example of evolution. Let us consider Darwin's mechanisms of modification.

Mutation

Mutation is a change in an organism's DNA, which results in a change in the way the organism develops, looks or acts. Some mutations can be useful for helping an organism to survive in its environment, but others can hamper this survival.

An important point here though, is that this usefulness or otherwise did not cause the mutation to occur. The mutation did not occur because it was useful. A mutation is not the result of some plan or strategy for survival. Nor is a mutation the result of the organism striving to improve itself. Mutations are random.

Germ Line Mutations

Mutations that are relevant to evolution are those that occur in reproductive cells—the ovum and sperm—and hence can be passed on to offspring. But even when a modification is inheritable, it may still be irrelevant to evolution. Some changes to DNA can be passed on but have no discernible effect on the organism's appearance and behaviour. The DNA that has changed in these cases is called 'junk DNA'.

Other mutations, however, do change looks and behaviour. The breeding of show animals for curly whiskers and soft fur are two of many examples of how such a mutation can be cultivated.

Beyond whiskers and fur, however, a mutation can significantly alter the characteristics, and crucially the survival prospects, of a whole species, for example in the case of insects that have developed a resistance to DDT.

Why Mutations Occur

When a cell divides, it makes a copy of its DNA, but sometimes this copy is not perfect. This difference between the DNA sequence of the original and the copy is a mutation. Mutations can also occur due to environmental effects like exposure to chemicals or radiation, naturally occurring or otherwise, which can cause DNA to break down. When DNA breaks down the cell repairs it, but this repair job may not be perfect. The repaired DNA may differ from the original, in which case it is a mutation.

Migration

Darwin's second mechanism of modification is simply the movement of genes from one population to another, which can occur when species move between colonies, or simply when pollen is blown to a new destination.

On an everyday level, when two individuals from different families produce offspring, 'genetic shuffling' occurs, bringing together a new combination of genes, so that the offspring may have the mother's eyes, but the father's hair, for example.

Genetic Drift

The concepts of mutation and migration explain how there comes to be variety in the genetic makeup of a population, but how is it that some of these variations survive and flourish, while others do not?

A first reason is simply that some individuals in a population will leave behind more descendants than others, hence their DNA is more likely to spread into the future. It is worth pointing out that this is not because these more fertile individuals are healthier or better specimens than their less fertile companions. This genetic drift is, rather, simply the result of chance.

Natural Selection

Critics of the theory of evolution like to depict evolution as a random or chance process. Darwin asserts that mutation is random. Natural selection, however, is not random. Natural selection is a principle or rule which explains why some mutations flourish and others do not. There is no room for chance in natural selection.

The idea of natural selection is that, given genetic variety in a population, the individuals in the population which are better equipped to survive, find a mate and reproduce, are the ones whose genes will be passed on to future generations. Suppose, for example, that green beetles are easier for birds to catch than brown ones. It is the brown ones, then, that will be able reproduce in greater numb and pass on their characteristics, including their colour. This is natural selection at work.

Fitness

In the context of natural selection, 'fit' individuals are not necessarily strong, fast or big. Fitness is simply a matter of being able to survive, find a mate and reproduce.

Indeed, this may be achieved with modifications that render an individual unfit in our normal sense of the word. The full tail-plume of a male peacock, for example, hampers the bird's mobility and may attract predators, but it does make the bird fit in an evolutionary sense, because it attracts more mates for reproduction.

Adaptations

Adaptations are characteristics of a population or species that improve its ability to survive and reproduce. An adaptation may be some particular behaviour, for example a means of evading predators or attracting mates, or an anatomical feature like the chameleon's ability to camouflage itself by changing its colouring.

Many of the strangest features of the natural world, and those which impress us most, are adaptations. But a mutation which begins its life as a useful adaptation may, due to changes in the environment or further modifications to the species, in time become useless.

The human appendix and the non-functional eyes of fish species that live in caves are examples of such modifications or vestigial structures, the remnants of past adaptations that no longer perform their original function.

Coevolution

Coevolution occurs when species that interact with each other affect each other's evolution. The interaction may be between predator and prey, parasite and host or between competitive or mutualistic species, such as plants and insects, where the insect feeds from the plant while also aiding its pollination.

To illustrate this evolutionary interaction, both red squirrels and crossbill birds in the American Rocky Mountains eat seeds from the cones of the lodgepole pine. In squirrel-inhabited areas, squirrel-resistant cones, with more weight and fewer seeds, survive to reproduce and have become the dominant kind. In crossbill-inhabited areas, crossbill-resistant cones, with large, thick scales, predominate. A consequence of this is that in the crossbill population in these areas with deeper, less curved bills, which can extract the seeds more effectively, have become dominant. Here, then, we see how adaptations in mutualistic species can coevolve.

FROM MICRO TO MACROEVOLUTION

Microevolution is evolution on a small scale, that is, within a single population or species. For example, global warming has led to changes in the DNA of populations of mosquitoes and insectpopulations have developed resistance to pesticides. These are changes within a branch of the tree of life, rather than being a new branch.

Such changes are generally measured in terms of the percentage of the population with the new characteristic—the percentage of brown beetles to green ones, for example. But if we accept the idea of the tree of life, much bigger changes than these must have occurred. How did they occur?

Species and Speciation

A population is a group of individuals that interbreed with each other. A species is all the individuals that can possibly interbreed. Two individuals may look as different as a poodle and a great dane, but if they can interbreed they are of the same species.

There are, however, borderline cases. Hooded and carrion crows, for example, look different and mostly mate within their own groups. In some areas, however, they do interbreed, producing hybrid offspring. If, however, these two populations were separated and acquired something as simple as different courtship behaviours, they might no longer be able to interbreed. They would then be independently evolving—different branches on the tree of life. This splitting of a branch is called 'speciation.'

Causes of Speciation

The most common way for speciation to begin is for two populations to be isolated from each other. This prevents interbreeding, and also subjects the populations to different environments, so that they are subject to different selective pressures.

If individuals from a long-haired species are removed to a warm climate, for example, they are likely over many generations to acquire a pelt more suited to their warm environment, as well as other possible modifications, such as different courtship rituals and mating habits, a lack of fit between sex organs (which is common in insect populations), and unviable or sterile offspring. Any of these modifications can rule out interbreeding.

Macroevolution

Macroevolution is evolution on a large scale: the appearance of a new species or a whole new branch of the tree of life, such as mammals.

Macroevolution happens too slowly to be observed, but geology, fossils and living organisms provide data allowing scientists to reconstruct the tree of life, creating a map of past macroevolutionary changes.

The mechanisms of modification—mutation, migration, genetic drift and natural selection—might seem too small-scale to generate whole new branches on the tree of life. Here it is important to remember the time-frame of the map, however ... much can happen over billions of years.

EVOLUTION AS SCIENCE

Evolution is a theory in the sense that it is a story about how all past and present life on our planet came to be as it was, and as it now is.

The basic idea is that all species evolved from shared ancestors. The work of scientists is to gather data and see whether that data fits with, or conflicts with, the idea of evolution.

In practice, however, scientists are more likely to simply assume the idea of evolution from the outset, in order to generate more specific questions. For example, if two populations of a species of beetle differ, a scientist may ask: how did this difference come about? The scientist will look at environmental differences and patterns of migration, and attempt to come up with a hypothesis: a story about what happened, which can then be tested. In this enquiry, the idea of evolution is simply assumed from the outset.

Studying beetles in this way makes the point that the story of evolution is in no way complete. It is a very general story, with many, many gaps yet to be filled in.

It is precisely these gaps which generate scientific enquiry, into beetle populations, for example. As scientists gather data to fill the gaps, it is always possible, in principle, that they could find data which simply cannot be made to fit with the story of evolution. This would imply that the theory of evolution is wrong. However, it is not immediately clear what sort of data or evidence could fail to fit with the story of evolution.

Scientists are constantly encountering things that they can't yet explain in evolutionary terms, but this does not lead them to doubt the theory. Rather, it compels them to investigate further, to come up with an explanation. So is it possible to challenge the validity of the theory of evolution? This is what ID aims to do. The next chapter outlines the theory of ID, to then consider the challenges it poses to the theory of evolution.

What is Intelligent Design?

The theory of ID devotes much time to criticising evolutionary theory. We shall consider these criticisms in the chapter entitled Arguments. Firstly though, let us consider the positive claims of ID, asking: what can this theory tell us about the origin and development of life on earth?

The critical question to keep in mind in this chapter is: does the complexity of earthly life justify the claim that there is a supernatural designer?

Paley's Watchmaker

The key idea of intelligent design (ID) arises from observing that the natural world is infinitely rich and complex. Ever since the time of the ancient Greeks, this complexity has led thinkers to conclude that the natural world must therefore be the handiwork of a cosmic creator. The best-known exponent of this view is English theologian William Paley.

In his 1802 book, *Natural Theology*, Paley wrote, 'If we find a pocket watch in a field, we immediately infer that it was produced not by natural processes acting blindly, but rather by a designing human intellect. Likewise, he reasoned, the natural world contains abundant evidence of a supernatural designer.

A New Creationism?

Critics of ID suggest that it only emerged as a theory after the 1987 Supreme Court ban on teaching Creation Science, as a way of smuggling God back into the science classroom.

Proponents respond that ID was already around, for example in Charles Thaxton's *The Mystery of Life's Origin* (1984). Thaxton aimed to investigate the possibility of design in nature without bringing in religious assumptions, and it is this avoidance of religious assumptions that distinguishes ID from Creationism.

ID simply argues that an intelligent cause is the best explanation of certain features of the natural world. This lack of religious assumptions is made clear by the fact that ID does not rule out the possibility that the intelligent designer may in fact be a hyper-intelligent race of aliens from another galaxy!

DETECTING DESIGN

Paley infers that a pocket watch is designed because of the complexity of its construction. This reasoning may be intuitively obvious, but the argument behind Paley's hunch can be spelled out.

Things in the world happen either by chance, because the laws of nature compel them to occur or because someone makes them happen.

How do we know which things have been brought about by an intelligent being, which things have been 'designed' in this sense?

In his book *The Design Inference*, CSC philosopher and mathematician, William Dembski, suggests that we can tell that something has been designed if it exhibits a property he calls 'specified complexity'.

Specified Complexity

Something exhibits specified complexity if it has a function, structure or purpose over and above the significance of its parts. Another way of saying this is that the thing needs to match some independently given pattern.

For example, the sequence of symbols 'DESIGN' conveys a recognisable meaning that is independent of the significance of the letters which comprise it: it matches the pattern we use in English for a word in our language.

The sequence 'ESDIGN,' however, does not. This is to say that while both sequences are complex, only the first is an example of specified complexity. Does this mean that the sequence has been designed?

Necessity, Chance and Design

In order to establish that a pattern or sequence exhibiting specified complexity has been designed, it is important to show, firstly, that it is not simply the result of a law of nature.

For example, the colours of a rainbow are a pattern, but this pattern cannot be otherwise. The same is so with the elaborate structure of a salt crystal. These patterns are simply a consequence of laws of nature.

In the case of DESIGN/ESDIGN, however, there is no law of nature compelling the letters to be in one order rather than another.

Secondly, then, could this ordering of letters have occurred simply by chance? A random throw of scrabble pieces could conceivably yield a six-letter sequence like 'DESIGN', because it is not very complex.

Choose a more complex pattern however, such as Hamlet's 'To be or not to be' soliloquy, and this possibility of random ordering becomes most unlikely, in which case we can confidently say that an intelligent being created it: it was designed.

Design in Nature

The whole point of this abstract argument is to establish design in nature, and here examples abound. For example, the eyes of all manner of species, from humans to squid, have a familiar, camera-like design, with recognizable parts: a pinhole opening for light, a lens and a surface upon which to project an image. These parts are arranged just as a human engineer would arrange them, which is to say they conform to an independently given pattern and so qualify as examples of specified complexity.

At a more fundamental level of the natural world, every living cell has a DNA molecule, which is a complex structure that provides 'instructions' for the workings of the cell. Each protein or 'letter' of the DNA is meaningless on its own: it is only as a chain or sequence that these elements function. Hence DNA exhibits specified complexity.

The variety of DNA sequences shows that no single sequence is a necessary consequence of natural law and the complexity of the sequence suggests that a random ordering is most unlikely. It seems, then, that this building block of life is designed.

Irreducible Complexity

Another way to detect design in nature is to look for biological structures that are 'irreducibly complex'. CSC biochemist, Michael Behe, explains in his book *Darwin's Black Box*, that a structure is irreducibly complex if it needs all its parts in order to function.

Consider, for example, a mousetrap. If you remove one part of the trap it doesn't become slightly worse at killing mice—it ceases to function at all.

So too within living organisms—the bacterial flagellum, for example, is a tiny outboard motor that bacteria cells use to propel themselves. The flagellum is comprised of numerous proteins, all precisely arranged, and if any one of them is removed the flagellum stops spinning.

The point here is that such a complex system could not have been formed slowly over time, because it can only perform its function as a complete unit. Such a system could not then have evolved. Rather, it can only have been designed.

ID AND SCIENTIFIC ENQUIRY

Much ID literature is devoted to pointing out perceived flaws in Darwin's theory of evolution.

These criticisms—which are discussed in chapter entitled Aguments—imply that an alternative theory is necessary, but they are not themselves part of an alternative explanatory theory.

They do not explain the nature of anything in the natural world. In contrast to this, the ideas discussed so far in this chapter, firstly, explain how to discern intelligently designed features of the natural world.

Secondly, they also explain the nature of these features themselves, insofar as they suggest that these features were designed.

What does this idea imply for a practicing scientist?

How should a scientist who accepts this idea proceed in his or her work?

Removing Origins from Science

In their book, *Intelligent Design: The Scientific Alternative to Evolution*, William Harris and John Calvert argue that, once the scientific community accepts ID, scientists can stop wasting time speculating about where life came from, and focus on the more interesting question of how life works.

The first claim here is that ID will close off fruitless and dubious avenues of scientific enquiry. Speculation about how life began is just that, speculation.

The origin of life happened so long ago that is not something that we can gather concrete evidence about in order to prove or disprove any theory about it. As such, although the question may be very interesting, it is not something scientific study can say a lot about.

Changes in the Science Lab

If scientists abandon the idea that species share a common ancestry, however, how much biological science is there left to study? The second claim of Harris and Calvert is that scientists can investigate how life works. This suggests that there is still quite a lot that can be studied.

The authors list some of the topics compatible with ID:
– 'Let us try to determine just how 'plastic' the genome is.
– What are the natural limits of variability and how far can those limits be extended by intelligent manipulation of genes?
– Can we turn a squirrel into a chipmunk by gene insertion/deletion?
– Can we cure genetic diseases?'

The general suggestion here is that genetics and biology can get along quite well without the idea of shared ancestry. ID can be accepted and much of current scientific practice can proceed relatively intact.

New Predictions

With this view scientists can accept ID and much of current scientific enquiry can continue relatively intact. Beyond this, does the theory of ID make predictions that open up new areas of enquiry?

The idea that parts of nature were purposefully designed does prompt a different attitude towards them.

In the previous chapter we looked at the idea of junk DNA: parts of a gene that have no discernible affect on the appearance and behaviour of the organism. If a gene was purposefully designed, however, this DNA must in fact have a purpose.

Although this is a very general claim, it can be called a 'prediction' of ID. This prediction could motivate scientists to analyse DNA with greater vigour, in order to discern the purpose of apparent 'junk'.

Reverse Engineering

Another area in which ID is assumed is in the practice known as 'reverse engineering'. Biochemists take apart biochemical elements in order to see how they were put together in the first place—how they were 'engineered.'

They are looking for 'design decisions' built into the architecture of the biochemical system which will help them understand how the system operates.

The assumption that they will find 'design decisions' is in fact a prediction which follows from the idea that these systems were, in the first place, designed.

ID, Evolution and Faith

Given that ID may be integrated into scientific practice in this way, the theory of ID has implications beyond science. Most obviously, the idea of an intelligent designer is compatible with religious faith, with belief in a creating god.

Now the theory of evolution does not seem to have religious implications, but in fact adherents of ID proclaim that the theory of evolution also entails a certain faith—that the theory of evolution implies a belief in the value of science, a view that the world is solely material. Importantly, this belief is not itself scientific: it is not based on evidence. Rather, it too is an article of faith.

Before considering arguments for and against these competing theories, we need to see how the argument between them, and more generally science as such, fits into the larger world of human concerns.

Faith in Science: What is at Stake?

In the opening chapter we acknowledged ID's claim—that the theory of evolution is part of a larger culture of 'materialism' in science—which proponents of ID say is damaging to modern society and human values.

In this chapter we analyse this modern materialism. We look at its history, and discuss the loss of ethical guidance that has accompanied modern science. This loss of ethical guidance is a social problem to which both ID and evolutionary theory propose a response.

In this chapter we sketch these responses, in order to give a context to the debate between ID and evolutionary theory.

THE RISE OF MATERIALISM

Darwin and Religion

Darwin delayed publishing his ideas for 21 years because he was concerned about the effect his ideas would have on society at large and specifically in relation to religious thought.

Since recorded history began, religion has had answers for questions about life, the universe and everything. As science developed, it proposed alternative answers, which conflicted with the religious account of how the world works.

Darwin was aware that his work was perhaps the final step in this challenge to religious explanation.

Heliocentrism and the Bible

One of the first known signposts of this conflict—between science and religion—is Galileo's claim (1616) that the earth revolves around the sun (that is, the earth and other planets are heliocentric). This contradicted the accepted 'truth' that the earth was the static centre of the universe, as a literal reading of the Bible implies. Prior to this, the stars were thought to be God's domain—not subject to earthly laws.

In Galileo's scheme, our earth itself became a star! No one, however, knew how or why the planets stuck to their courses as they did, until Isaac Newton proposed the idea of gravity (1666). Here a law which applied in the human sphere (things fall downwards), was shown to apply in the heavens as well.

A testable scientific explanation of planetary motion replaced a religious explanation, and the traditional boundary between the human earthly sphere and the divine heavenly sphere was thrown into doubt.

The Enlightenment

Sir Isaac Newton is a fitting representative of the 18th century Enlightenment. On the one hand Newton propounded a thoroughly scientific theory of the workings of the natural world. He did not appeal to the supernatural, but concentrated on observing regularities in nature and establishing the laws that underlay these natural features.

At the same time, however, Newton wrote voluminously on the influence of angels in human life, thus reflecting the Enlightenment concern with human morality and piety.

These concerns of the human soul were understood to be something quite different from regularities in the natural world.

Although religion could not explain regularities in the natural world, it still had an important role to play: offering guidance in human affairs.

Darwin and the Modern Era

By the middle of the 19th century, the scientific method of analysing nature was being applied to humans as well as the external world. The argument of Modernism is that humans are no different to other parts of nature: human behaviours and activities are simply the result of certain general laws of nature. Humans have no higher nature or calling.

Sigmund Freud, the father of psychoanalysis, argued that humans are not free, rational beings. Rather, human rationality is simply the result of sexual drives being channelled and directed by certain laws of the unconscious. Similarly, Karl Marx argued that we do not choose the way we live together. Rather, different forms of human society are simply the result of certain economic laws, which have directed our social history.

Darwin's theory appeared to be the final nail in the coffin of traditional religious explanation. He argued that humans were not different in kind to other animals. The human race has simply evolved skills that allowed it to dominate nature.

This notion that human beings were driven by the same impulses as 'lower animals' was difficult to reconcile with the Biblical idea that humans alone were created in the image of God.

Freud, Marx and Darwin differed in fundamental ways, but had one thing in common: each of them attempted to explain humans not as the result of a god's creation, but as the result of laws that govern forces and matter in the natural world.

Modern Materialism or Methodological Naturalism

The 19th century English naturalist Alfred Wallace argued that evolution applied to everything except the human mind and spirit, which was God-given. Darwin, however, held that even human consciousness could be explained as the result of the physical, material forces of mutation and natural selection.

So, the human mind and spirit was not essentially different to the natural world, but was simply a part of it. Moreover, mutation is random: it is not guided by some superior mind or spirit such as a god. It is this total rejection of the ideas of human or divine spirit in scientific explanations of the world that defines modern materialism.

In science, this viewpoint is also called 'methodological naturalism.' This name makes clear that this stance is actually just a method: a way of doing science.

A scientist may well believe in a god and the human soul. It is just that he or she will not use these ideas in scientific explanations of the world. Science and religion are separate. (See the final chapter: Where to Stand? for more on this position.)

MATERIALISM AS AMORAL

DNA and Ethics

Since Darwin's time, science and religion have largely gone their separate ways. The scientific question 'how does it work?' has been separated from the religious or ethical question 'what is the right thing to do?' This has happened at a time when new discoveries are constantly being made in science. Since the discovery of DNA, biological knowledge has increased exponentially, especially in the field of genetics.

Modern science can achieve fantastic, life-saving 'miracles', but it has also given us the ability to annihilate whole segments of the human population, as well as other species. This greater understanding of 'how things work' has made ethical questions all the more compelling and urgent: what is the right thing to do with these new discoveries and powers?

Since religion has been weeded out of science for methodological reasons, science has lost any ethical framework or guidelines for considering these questions.

Human Purpose and Responsibility

Some think that this lack of a moral compass in science is the direct result of Darwin's ideas. ID proponent Phillip Johnson, for example, suggests that if humans evolved by natural selection, it is not possible for life to have a purpose. This is to say, Darwin's theory provides no answer at all to the ethical question: what is the right thing for humans to do?

Theologian Jay Richards extends this idea, saying that if humans arose purely by chance, we are not responsible to any higher power, hence we can do whatever we want. The CSC authors, William Harris and John Calvert, follow this idea through:

– 'If life is an accident, why not alter it to suit our needs?
– If we can, why not make human clones?
– Why not abort unwanted children?
– Why not euthanise the useless aged?'

The implication here is that modern science, and humans living in the age of modern science, are confronted with too many choices and are desperately in need of moral guidance.

Science as an Amoral Religion

Science has sometimes been viewed as a religion because it assumes that there are 'laws' underlying the physical world that make the world function as it does. If they govern our world from outside, then these laws themselves must then be metaphysical, or super-natural, much like a god and, like a god, these laws are unknowable, yet are held to make the world as it is.

The theory of evolution in particular can also be considered as a religion in a different sense. Philosopher, Alvin Plantinga, suggests that evolution is like a religion because it has ethical implications. The theory of evolution provides us with 'a deep interpretation of ourselves, a way of telling us why we are here, where we come from, and where we are going.'

This interpretation is, again, essentially negative: we arose by chance and have no superior purpose. If science is a religion, it preaches that there is no purpose and value in life.

REINTRODUCE GOD, OR KEEP SCIENCE MATERIALISTIC?

Design Implies Purpose

The development of modern materialist science has resulted in a perceived loss of moral guidance, both for science itself and for people in modern society. There are two possible responses to this problem. The first is to attempt to reintroduce a moral or religious dimension back into science. Proponents of ID implicitly advocate this approach.

ID does not claim to provide this moral dimension or framework itself. All ID argues is that the idea of a designer is necessary in scientific explanations of the origins of life. But this claim—that life was designed—implies that life was designed for some purpose.

Science may not be able to help establish what the divine purpose of life is. This would seem to be a question that only religion or philosophy can help answer. The reintroduction of the idea of a designer into science, however, would at least confirm that there is a purpose to be discovered.

This divine purpose would provide a basis for scientific enquiry, and for distinguishing right and wrong, for answering the question 'what is the right thing to do?' in any situation. For 'the right thing to do' would be whatever most accords with a particular god's divine purpose.

Science and Religion

Answer Different Questions

A second approach to the lack of moral guidance in materialistic science is to simply look elsewhere for such guidance.
Author and Rhode Island biology professor, Kenneth Miller for
example, is a scientist and a devout Roman Catholic. He
believes that the theory of evolution can only explain how life
arose and how it diversified.

Why there is life at all is another question entirely, one that
Miller believes is outside the realm of science. Miller presumably turns to the Church for answers to the 'why?' questions
about life, and the Church presumably provides Miller with a
sense of purpose in his life and a basis for discriminating right
from wrong.

Miller doesn't ask science for moral guidance. Nor he does not
want science to attempt to provide such guidance, as this
would compromise the methodological materialism that,
historically, scientists have striven so hard to achieve.

Science, as a way of knowing, has been extremely successful.
Keeping religion out of scientific explanations has meant that
scientists are compelled to come up with purely materialistic
explanations of how the world works. This is what drives science,
and this is what Miller and others like him want to retain.

Which is Better Science?

This, then, is what is at stake in the debate between the theories of ID and evolution. In order to become more effective, science has become separated from religion, but has thereby lost an ethical dimension.

Should this ethical dimension be reintroduced into science, or be found independently of science?

It would seem that either option would provide the necessary ethical framework. If this is so, then the pertinent question to ask is simply: which option would result in better, more compelling scientific theories?

Let us now turn to the arguments for and against ID and evolution.

The Arguments

The debate between ID and evolutionary theory has its origins in the United States, but has taken place in journals, magazines and books throughout the world. In most cases the debate follows a regular and repeated pattern. It begins with a criticism of the theory of evolution, in which it is argued that evolution fails to explain some feature of the natural world. This may then be followed by an alternative ID-based account of the feature of the world that is at issue. In response to this, a proponent of evolution then attempts to rebut these arguments by demonstrating that the criticism and the alternative account are both flawed, hence the theory of evolution is still the best explanation.

The arguments that diverge from this pattern mostly concern the method that each theory employs. Here, what the theory can and cannot explain is not the issue. Rather, the issue is whether the theory is in fact a properly scientific theory at all. In this chapter we look firstly at arguments about what the theory of evolution can and cannot explain. As well as developing its own criticisms, ID also draws on more general criticism of evolution, by Creationists and other opponents of Darwin. Let us begin with these general criticisms of the theory of evolution, then turn to ID-specific criticisms. After that we will look at arguments about the method or status of each theory.

GENERAL CRITICISMS OF EVOLUTION

Intelligent Design

Evolution is Statistically Improbable

World-renowned astronomer Fred Hoyle famously said that evolution was as likely as a tornado blowing through a junkyard and assembling a Boeing 747 jumbo jet. His point is that plants and animals are complex biological systems and it seems totally improbable that these complex systems could have developed out of primitive cells in some primordial swamp, simply as a result of chance-based Darwinian mechanisms.

At a more fundamental level, even the 'simple' proteins necessary for life are complex. The odds of even one simple protein molecule forming by chance are one in 10 to the power of 113, and thousands of different proteins are needed to form life.

For further probability-based arguments, see page 100: Blind Evolution Cannot Generate Complexity (No Free Lunch), and page 102: The Universe is Finely-Tuned.

Evolution

Natural Selection is Not Random

Although gene mutation in a population is random, the mechanism of natural selection is not. It says that genetic variants that aid survival and reproduction are more likely to flourish. Hence complex biological systems do not come about simply by chance, but rather due to their precursor's ongoing fitness in their environment.

The same is so with proteins: their environment dictates the ways they can combine. And it is worth noting that there are innumerable possible proteins that can promote life. So while the appearance of one particular protein formation (the one that made us) may be unlikely, that some life-promoting protein or other will appear is much more likely.

Finally, when considering the likelihood of complex organisms evolving, the time-frame of evolution must be taken into account. Today's complex organisms are the result of over three billion years of selecting random mutations.

Intelligent Design

The Origin of Life (Abiogenesis)
is Pure Speculation

The idea of evolution implies that at some time in the primordial past living cells somehow spontaneously evolved from non-living matter. Attempts have been made to explain how this spontaneous generation occurred, but they have failed.

Over 80 years ago, in 1924, Russian biochemist and evolutionist, Alexander, Oparin suggested which chemicals must have been present in earth's atmosphere for amino acids to form. In 1953 the University of Chicago graduate student, Stanley Miller, attempted to replicate these conditions.

Miller's experiment did produce organic compounds including amino acids, and other simulated environments even yielded nitrogenous bases, which are a component of the building blocks for DNA. However, in all these atmospheres oxygen was excluded as it would prohibit acid formation.

Earth's atmosphere, however, includes oxygen, and even the oldest rocks include oxides, which is evidence that these rocks were formed in the presence of oxygen. Hence these experiments tell us nothing about the actual oxygenated origin of life on earth, and the idea of evolution lacks a starting point.

Evolution

Evolution Does Not Depend on Abiogenesis

The fact that we have oxygen in the atmosphere today is mainly the result of photosynthesis. Before photosynthetic plants and bacteria evolved, about 2.5 billion years ago, there would have been little oxygen, as it had no source. The atmosphere was most likely high in hydrogen, which facilitates the formation of organic molecules. Hence Miller's experiments *are* relevant to the origin of earthly life.

There is a great deal about abiogenesis that is unknown, but investigating the unknown is what science does. Speculation is part of the process. As long as a speculative hypothesis can be tested, it is scientific.

With regard to the theory of evolution, however, it is true that evolution implies that life must have had an origin, but whether the origin of the first replicating molecule is ever fully explained has nothing to do with what happened subsequently. Evolutionary theory was not proposed to account for the origin of all living beings, only the process of change once life exists.

Intelligent Design

New Species Suddenly Appeared in the Cambrian Explosion

According to the fossil record, something quite strange happened half a billion years ago, in the Cambrian period. Before this time almost all life was microscopic, except for some soft-bodied organisms.

At the start of the Cambrian period, however, animals burst forth—ocean creatures acquired the ability to grow hard shells, and a broad range of new body plans emerged within the geologically short span of ten million years.

Nearly all the animal types (phyla) in existence today, as well as some no longer with us, first appeared in the fossil record at this time. These animal groups appeared abruptly and fully formed, and are ancestors of virtually all the creatures that swim, fly, and crawl today.

Palaeontologists have proposed many theories to explain this explosion in the fossil record, but have agreed on none.

This explosion is a problem for Darwinian evolution, as Darwin believed that all living things are modified descendants of one or a few original forms, hence life developed in the structure of a tree.

If all the major groups of animals emerged at once, however, this structure is closer to a lawn or tangled thicket, and suggests the possibility of divine, creative intervention at this time. Indeed, this Cambrian appearance of complex animal types has a 'top down' pattern.

Evolutionists would expect to see increasing variation over time, until significant new body plans are achieved. Instead, the fossil record demonstrates the abrupt appearance of radical new designs first, followed by minor variations there-after.

This is precisely the pattern one would expect from an intelligent designer.

Evolution

The Cambrian Explosion is Consistent with the Tree of Life

The Cambrian Explosion was not the origin of all complex life. There is evidence of multi-cellular life before the Cambrian period, including sponges, ancestors of jellyfish and at least six other types of organism (phyla).

Other soft-body types appear much later and all plants post-date the Cambrian. Similarly, almost none of the animal groups we now recognise—mammals, reptiles, birds, insects, and spiders—appeared in the Cambrian.

So the Cambrian Explosion does not show all groups appearing together fully formed.

So why did this explosion in the fossil record occur? During the millions of years in question earth was coming out of an ice age, and it is likely that the amount of oxygen in the atmosphere was increasing, opening new niches for life to evolve into and allowing new species to flourish.

Beyond this, the appearance of active predators spurred the coevolution of defensive shells, teeth, and other hard structures, which fossilise more easily, ensuring a richer fossil record than from the Precambrian period.

The Precambrian fossils that have been found, however, do confirm a branching pattern of development. Bacteria appear well before multi-cellular organisms, for example, and plants diverge from a common ancestor before fungi diverge from animals. Moreover, if the number of cell types in an organism is a measure of its complexity, organisms have become more complex more or less constantly since the beginning of the Cambrian. Finally, explosions of new species have occurred in other periods, such as the Ordovician, and while these explosions pose interesting questions about environmental influences, they are not sufficient to cast doubt on the basic idea of evolution: descent with modification.

Intelligent Design

There are no Transitional Forms as Evidence for Speciation/Macroevolution

Darwin asserted that 'living beings evolved gradually' in which case the fossil record should present us with a host of 'transitional form'. That is, organisms which are no longer of one species, but neither have they fully evolved into a recognisable new species. For example, if mammals evolved from reptiles (which they didn't), shouldn't it be possible to find at least one example of the half-reptile/half-mammal that was an intermediary in this hypothetical transition?

There is no evidence of one species morphing into another. It is true that there is variety within a species—variations in size, for example. And it is true that an already existing trait of a species can be selected and come to predominate (microevolution). But there is no evidence that a species can acquire altogether new traits. Nor is there evidence that a species can acquire enough new traits that it transforms into a new species (speciation, which is a part of macroevolution).

As one evolutionary palaeontologist says of the fossil evidence, 'When a major group of organisms arises and first appears in the record it seems to come fully equipped with a suite of new characters not seen in related, putatively ancestral groups. These radical changes in morphology and function appear to arise very quickly.'

Evolution

There are Gaps, but There are also Transitional Forms

The distinctness of species and animal forms is partly an illusion created by the way we divide them up. Quite different types, including transitional forms between them, are lumped together under one name, and quite similar things are distinguished. The line between dinosaurs and birds, for example, is arbitrary, as many fossils of intermediates have recently been discovered, including dinosaurs with feathers, with an elongated hand/wing, with a keeled chest bone and V-shaped furcula (wishbone), and without teeth.

Similarly in human ancestry, the leg and pelvis bones of *Australopithecus*, the most primitive human ancestor, show that it walked upright, but the bony ridge on its forearm shows that it had previously walked on its knuckles: it is a transitional form.

There are gaps in the fossil record, but this is to be expected, as fossilisation is not a common event, nor is finding fossils. Given this, transitional forms have been found, and these are evidence for both speciation and macroevolution. Also, although large scale macroevolution occurred too long ago to have been observed, mutation and speciation have been observed. Some bacteria have acquired the ability to digest nylon, unicellular green algae have mutated and become multicellular and several altogether new species of plants have arisen due to multiplication of their chromosome count.

Intelligent Design

Similarity of Parts: Homology, Analogy or Design?

A horse's leg, a bat's wing, a porpoise's flipper and a human hand all contain bones that are structurally similar. Before Darwin, biologists called this homology, and considered it evidence of a common design. Darwin, however, attributed such homology to a common ancestor. A first problem with this is that homology is defined as the result of common ancestry, but is then put forward as evidence of common ancestry.

This is a circular argument. It amounts to saying that two features are homologous because they come from a common ancestor, and they come from a common ancestor because they're homologous.

A second problem arises when evolutionists attempt to distinguish homology from analogy. To explain what is at issue here, consider the amazing similarities that exist between the organs of very different species.

For example, humans and squid are quite unrelated species. No strong evolutionary link between them has ever been proposed, yet the eyes of both are strikingly similar, both in their structure and in their function. Because no common ancestor with eyes is apparent, however, evolutionists say that these organs are not homologous, rather they are analogous: very similar, but with no evolutionary link.

Once again, the reasoning here is circular. This distinction between homology and analogy is not made on the basis of the observable evidence, for what is observed is simply that squid and human eyes are very similar. The only reason for distinguishing between similar features which are homologies, and similar features which are analogies, is that you have already assumed that the theory of evolution is correct. For the theory simply stipulates that this difference must obtain.

Before pointing to a much simpler answer to this problem, it is worth noting that this development of similar organs in otherwise quite different species poses a further problem for the theory of evolution.

The eye is a complex organ. How is it that a host of independently evolving species all managed, independently of each other and by random mutation, to develop this same organ, with the same camera-like design of a pinhole opening for light, a lens, and a surface upon which to project an image?

Darwin was right when he said that to suppose that the eye had evolved was 'absurd in the highest possible degree' and to suppose that it evolved repeatedly in different species is even more so. A simpler explanation of why quite different species have similar organs is that the eye, for example, is a useful design feature, which has been repeatedly employed by a designer in creating the huge variety of natural species.

Evolution

Shared Derived Characters, Adaptations and Darwin on the Eye

Homology alone is not sufficient to establish shared ancestry. A biologist does indeed begin by looking for similarities, but these are not simple likenesses. Rather, the biologist is looking for shared derived characters: characteristics that a number of species share, but which are not apparent in other contemporary species or in earlier species. For example, although they all possess a bony skeleton, amphibians, turtles, lizards, crocodiles, birds and mammals all have four limbs, whereas ray-finned fishes do not.

Having four limbs is thus a shared derived character: no prior or contemporary species has it, only this group and possible descendants thereof. But defining a shared derived character is only the first step towards establishing shared ancestry. Similar categorisation must be undertaken with all the other defining characteristics of these beasts and their ancestors.

On the basis of this, and knowledge of when and where each species existed, a detailed mapping must be made of the species' relation to each other and to their forebears, which shows their place in the larger tree of life. It is only at this stage in the analysis that shared ancestry is taken as established.

In this mapping process two species may share one characteristic but differ in many others. Bats and birds, for example, both have wings, although in all other respects bats share more shared derived characters with mice than with birds. In fact closer inspection reveals that these wings are quite different in structure: a bat's wings are flaps of skin stretched between the bones of the fingers and forelimbs, whereas bird wings consist of feathers extending along the length of their forelimbs. It is on the basis of these structural differences, and the larger mapping of shared derived characters, that biologists assert that bat's and bird's wings evolved independently, and their similarity is just an analogy.

So how did bats and birds, evolving independently, both come to have wings? Mutation is random, but natural selection is not, and in the environment in which the ancestors of bats and birds developed, the ability to fly was a most definitely an advantageous adaptation. Similarly, eyes have independently evolved in numerous species as an advantageous adaptation in response to environmental pressures.

After writing that the evolution of eyes seems absurd, Darwin listed seven intermediate stages via which eyes may have evolved: from photosensitive cells to aggregates of pigment cells, to a covered optic nerve, and so forth. Eyes do not fossilise well so this developmental path cannot be proven, but it is note-worthy that each of these stages of development can be found in animals living today.

ID-SPECIFIC ARGUMENTS

Intelligent Design

Irreducible Complexity

The idea of irreducible complexity is the idea that some organisms, or parts of organisms, are so complex that they could not have been formed one step at a time, as Darwin claims they were. In his book *Darwin's Black Box*, biochemist Michael Behe explains explains this idea using the everyday example of a mousetrap.

A mousetrap is made of five parts, and all the parts must be in place for the mousetrap to work. If you remove one part of the trap it doesn't become slightly worse at killing mice—it ceases to function at all. So the mousetrap couldn't have slowly developed, once piece at a time, from a primitive mousetrap to a modern one. Until all five pieces were there, no mice were caught.

In addition, parts of plants and animals are complex in this same way—the bacterial flagellum, for example, is a tiny outboard motor that bacteria cells use to propel themselves. The flagellum is comprised of numerous proteins, all precisely arranged, and if any one of them is removed the flagellum stops spinning, or cannot even be built by the cell.

Another example is the eukaryotic cell, which is comprised of little compartments that do different jobs, like digesting nutrients and excreting waste. Proteins enter these compartments with the help of 'signal' chemicals, which turn compartment reactions on and off. This regulated flow of traffic in a cell requires that all parts function together, or the system breaks down. This, then, is another example of an irreducibly complex system.

Biology studies the workings of these cells, but can say little about how they could have evolved by natural selection, because natural selection can only choose among systems that are already working. These complex systems could not have evolved one step at a time, because until all the parts were in place, these systems simply could not function at all. The complexity of these systems suggests that they were, rather, designed.

Evolution

Complexity Reduced

Co-option is an important principle in biology. The idea is that a feature of an organism can be selected for one use, but once established can be co-opted for another use. Kenneth Miller demonstrates this using the example of the mousetrap: the base alone can serve as a paperweight; combined with the spring it becomes an effective mini clipboard, and so on.

The point is that bits and pieces of this supposedly irreducible system may initially have one use and then, in combination with another piece, be co-opted for different use. In the natural world a series of co-options like this could explain how a complex system gradually evolved.

In the bacterial flagellum, for example, it is true that the flagella need all their proteins to operate, but these proteins can also perform other functions. For instance, the bacterium can use these proteins to inject poisons into other cells. These proteins may, then, have been favoured by natural selection for quite a different reason than for the job they do now. This is why they were already in living cells and were able to be co-opted to their new role later on.

Although the bacterial flagellum is indeed a complex system, it could have evolved one step at a time, with its parts performing different functions as it evolved.

Similarly, in the case of the eukaryotic cell, which uses so called 'signal' chemicals to direct the traffic flow of proteins. Some of these proteins can direct their own secretion so that no transport mechanism is necessary. Hence they could have already been in place, ready to be co-opted into their new use.

Beyond this, many of the proteins involved in transport in eukaryote cells have molecular ancestors in bacteria. These ancestor molecules already serve in much simpler systems. Hence they, if not also the system they operate, could have been co-opted in the evolution of the eukaryotic cell.

Intelligent Design

Specified Complexity

Mathematician William Dembski's idea of specified complexity is that certain features of the natural world must have been designed because, firstly, they are too complex to have occurred by chance, and, secondly, they conform to an independently given pattern. For example, the eyes of various species, from humans to octopus, are complex and, as we have seen, have a familiar, camera-like design, with recognizable parts. This structure of the eye is too complex to have occurred by chance. Moreover, these parts are arranged just as a human engineer would arrange them, which is to say they conform to an independently given pattern and so qualify as examples of specified complexity, and must therefore have been designed.

At a more fundamental level of natural world, every living cell has a DNA molecule, which is a complex structure that provides 'instructions' for the workings of the cell.

These instructions are provided by the complex pattern or sequence of the DNA, which in this regard functions like a word or sentence in a language. A word or sentence conveys information because it is part of a larger language: it conforms to the larger pattern of this language, which is independent of the actual words used. So too there is a language of DNA, comprised of all possible DNA combinations, and it is by conforming to this independently given pattern that a complex strand of DNA does its work. Hence DNA exhibits specified complexity, and must therefore have been designed.

Evolution

Shared Environments Yield Shared Patterns

Organisms evolve and some become more complex. They exhibit complex patterns, but what does it mean to ask whether they conform to an independently given pattern?

It is true that numerous species have eyes, and these eyes have a similar pattern or structure. But this can be explained as a result of natural selection.

As each species' eyes slowly evolved, each modification gave them an advantage in their environment. Being able to discern light, the direction it is coming from, shadows and movement, for example, aided survival and hence enabled procreation.

The similarity in eye structure across species is simply due to the fact that in each case this evolution is responding to the same environmental problem: being able to discern predators and prey. (Note that bat sonar is an alternative response to this problem, albeit in lightless environments.)

This is an explanation of how species can develop organs that have a complex structure, and which are similar to each other. But this explanation does not imply that these structures are conforming to some independent blueprint. Their similarity is simply due to the shared environment and environmental problems, in which these structures evolved.

This same argument can be made in relation to DNA. Although scientists are still working out how DNA transmits its instructions, they do know that DNA mutates randomly, producing different characteristics in an organism. Some of these characteristics are selected over others due to their advantage, which means that some patterns of DNA are selected over others.

Again, there is no reason to postulate an independent blueprint.

Intelligent Design

Intelligence
Can be Detected

Many well-accepted and uncontroversial scientific disciplines study physical evidence and determine that it is the handiwork of an intelligent agent.

Forensic science determines whether a death is accidental or intentional; cryptanalysis determines whether patterns of characters are random or an intentionally coded message; archaeology determines whether an artefact was intentionally created; and NASA's Search for Extraterrestrial Intelligence program (SETI) scans the heavens with radio telescopes searching for patterns of signals that could only have come from intelligent sources.

ID's inference of intelligent design is no different in method to these accepted sciences.

Evolution

Natural and Supernatural Agency are not Comparable

ID is radically different from these listed sciences for the reason that all, bar one, are attempting to establish human handiwork. This means that an investigator can make assumptions about human psychology, motivations and intentions, as well as looking for concrete evidence of human involvement, such as fingerprints.

If ID's designer designed all of nature, this designer must itself be super-natural, in which case we can know nothing about its psychological motivations, nor what its fingerprint would look like.

This is a very different situation for inferring intelligence. It is noteworthy in this connection that the SETI program makes no assumptions about extraterrestrial intelligence. It is not looking for complexly patterned signals, for example, but simply for ones that appear artificial.

Intelligent Design

Blind Evolution Cannot Generate Complexity (No Free Lunch)

Evolution can be understood as a search formula or algorithm. In the face of a problem, a new disease for example, the Darwinian algorithm of random mutation plus natural selection searches for a solution, such as disease resistance.

In his book *No Free Lunch*, philosopher and mathematician William Dembski points out that the No Free Lunch theorems (NFL) in physics prove the surprising fact that, when you specify a target and an algorithm for locating the target, no algorithm performs any better than a simple 'blind' search would.

This implies that evolution is no better than a blind search for solutions to evolutionary problems, in which case the only explanation for natural complexity is that it was designed.

Evolution

NFL Theorems Don't Apply
to Evolution

NFL theorems only apply to situations where the target can be defined independently of the algorithm used for achieving the target. In the case of evolution, the target of 'fitness' would need to be defined independently of the idea of random mutation plus natural selection (the algorithm for achieving fitness).

Since populations evolve by random mutation and natural selection they affect the environment and each other. As they alter the environment in which fitness can occur, they are effectively altering the definition of fitness itself, for what constitutes fitness in one environment is not the same as in another.

Thus NFL theorems do not apply to evolution.

Intelligent Design

The Universe is Finely-tuned

We live in a universe that is finely-tuned for the existence of life. The universe is held together by precisely determined physical laws that define the strength of gravity, nuclear forces, the mass of the electron, the charge of the proton, the chemical mix of our planet's atmosphere and so forth.

All these fundamental constants are precisely balanced. Were they even slightly different, not only would life not exist, nothing of any significance would exist. Too many variables are involved in this fine-tuning for it to have occurred fortuitously, or by chance. Rather, it could only have been designed.

Evolution

Life is Finely-tuned

The universe is perfectly suited to the life that has evolved in it and there is a good reason for this: life evolves in response to its environment, which is to say that the life forms which survive and flourish are those that are most finely-tuned to their environment, rather than vice versa.

Of course, if some basic features of the universe were different, then any life that evolved would be different too, but that is only to be expected for the reason just given. Beyond this, there may be some fundamental conditions that would make any form of life impossible, but we don't know all the life forms that are possible, hence we can't say what conditions would rule them all out.

Intelligent Design

Evolution Violates the Second Law of Thermodynamics

This law ostensibly concerns the conservation of heat, but it can be restated by saying that when energy is exerted in the world, some of this energy will be lost or dispersed. For example, some of the energy spent spinning a wheel will be soaked up by air and axle friction, hence less energy is given out by the wheel than was put into spinning it. This lost energy is called entropy, and the second law of thermodynamics can be stated by saying that as one thing happens and makes another happen in turn, entropy (lost or dispersed energy) can only increase.

Entropy is spent energy that loses its direction. As such, it can be understood as chaos or disorder within a system. Construed in these terms, the second law says that a system will, over time, tend to become less ordered—more chaotic. The theory of evolution, however, proposes the opposite: that a disordered and chaotic system produced a more complex and orderly system; the disordered primordial swamp gave rise to ordered complex organisms. This claim violates the second law of thermodynamics.

Henry Morris suggests that a more ordered system could only have come about with the aid of some kind of code or program, to direct the ordering process. These instructions could not have come from within the system; they could only have been issued by an intelligent designer.

Evolution

Evolution is Compatible with the Second Law of Thermodynamics

The second law of thermodynamics states that overall entropy (disorder) of the universe increases, but this does not mean that some parts of a system cannot become more orderly and complex. All that matters is that this increase in order be balanced by an equal or greater decrease in order elsewhere.

The most obvious example of this is the evolution of plant and animal species on earth, which was and is made possible by sunlight and heat from the sun. This flow of energy from the sun entails a rise in entropy in the sun. The heat and illumination from the sun makes life on earth possible; it makes it possible for seeds to grow into trees, and for simple organisms to evolve into more complex species.

Order increases here on earth, but the sun can only accomplish this by gradually running down. Billions of years from now, the sun's entropy will win out and the sun will fail. There is no need, in this explanation, to propose a designer or a set of instructions.

METHOD OR STATUS?

Intelligent Design

Evolution is Just a Theory

In *Implications of Evolution*, English biology professor Gerald Kerkut defines the general theory of evolution as: 'the theory that all the living forms in the world have arisen from a single source, which itself came from an inorganic form.'

In Kerkut's view, the evidence which supports this claim is not sufficiently strong to allow us to consider it as anything more than a working hypothesis. The theory of evolution makes claims about events that happened billions of years ago. No one was around to see these events, nor is it possible to repeat these events and verify that the claims of evolutionary theory are correct.

On the level of common sense, some of the central claims of evolutionary theory seem implausible. If we breed cats for a thousand generations, won't they still just be cats?

We have been aware of the idea of evolution for nearly two hundred years, but no one in that time, or before, has ever seen a species turn into something else. Indeed, if a cat was born with two tails we would call it a freak of nature, not an example of evolution at work. Of course we have seen minor changes within a species, beetles becoming darker in colour, for example, but the general characteristics of the species remain unchanged.

Finally, it is true that the fossilised bones of different horses, for example, have been found at different depths in the earth's strata. But this does not imply that the deeper one is more primitive or that it evolved into the shallower one, which is therefore 'more advanced'. Couldn't it simply be that they are different sorts of horses? Cases like this do not demonstrate evolution; rather evolution is simply assumed in the way we have been taught to describe them.

Evolution

Evolution is More that Just a Theory

In everyday talk a 'theory' is a guess or possibility. In science a theory is an explanation with testable predictions. It is more than just a tentative guess.

It is true that the time-frame involved means that we cannot see large scale evolutionary change in action, but nor can physicists see subatomic particles directly. They verify the particles' existence by the telltale tracks they leave in cloud chambers. This absence of direct observation does not make the conclusions of the physicists less certain.

The theory of evolution is supported by observations in genetics, anatomy, ecology, animal behaviour, palaeontology, and elsewhere. To challenge the theory, it must be shown that this evidence is wrong, irrelevant, or that it fits another theory better.

Given that we cannot see large-scale evolutionary change in action, we have seen fossils which clearly imply such macroevolution: the transitional forms between dinosaurs and birds, for example. We have also seen speciation, albeit on a small scale: unicellular green algae becoming multi-cellular, for example, and new plant species arising due to multiplication of their chromosome count.

It is also true, as Kerkut says, that the theory of evolution implies that life must have originated from inorganic matter and hypotheses about this abiogenesis are highly speculative.

Whether the origin of the first replicating molecule is ever fully explained has nothing to do with what happened subsequently. Evolutionary theory was not proposed to account for the origins of all living beings, only the process of change once life exists.

Evolution, as the theory of descent with modification, is well-supported by evidence. Eminent American science writer and palaeontologist, Stephen Jay Gould, argues that evolution is more than a well-supported theory: evolution is a fact and a theory.

Evolution is something that has happened and continues to happen in the world. The theory of natural selection is an explanation of the mechanisms by which this fact of evolution happens.

Intelligent Design

Refusing the Design Explanation
is Unscientific

Enquiry into how life came about is, as Phillip Johnson observes, a historical exercise. As such, it involves asking different sorts of questions than in other sciences. For example, when a physicist considers the force required to move an object, the terms in which the answer will be given are implied by the question (some degree of force). However, if we ask 'what happened to cause life to arise on earth?' the question does not dictate the terms of the answer. Perhaps life may have arisen due to chemical processes, but it is also conceivable that a designer could have played a role.

Such a designer has not been seen, nor have the subatomic particles that physicists make claims about, nor have the evolutionists' large-scale evolutionary changes. These latter are both unobservable. Their existence has been inferred because they are held to explain other evidence. The designer inferred by ID is no different.

Some scientists may reject the idea of a designer because they adhere to methodological naturalism: the idea that science must limit itself to studying only forces, matter, processes and events that occur in the natural world.

This view damages to science as a truth-seeking enterprise, for the following reason: In historical sciences, theories cannot be tested by predicting outcomes or repeating experiments. Rather, a theory is tested by comparing its explanatory power against competing theories, to determine the best explanation.

Scientists who adhere to the viewpoint of methodological naturalism refuse to consider the possibility of a designer. Hence they refuse to compare the explanatory power of ID with that of its rivals, compromising the process of theory selection and thereby science itself.

Evolution

The Design Explanation
is Not Scientifically Useful

To infer the existence of subatomic particles or large-scale evolutionary changes is to say that these things must exist in the natural world, even though they cannot be seen, because they help to explain other things that go on in the natural world.

To infer the existence of a supernatural designer is quite different: it is to say that something exists *beyond* our observable natural world which it nevertheless exerts an influence on.

Scientists generally confine their discussion to elements of nature (methodological naturalism). A supernatural element might be acceptable in a scientific explanation if it actually served to explain what goes on in the natural world in a scientifically useful way.

A feature of the natural world can be 'explained' by saying 'it's magic', but this is not scientifically useful. It does not imply any further predictions which can be tested, nor does it shed new light on any other aspect of the natural world.

Darwin's theory of evolution makes predictions that have been tested in numerous experiments and has shed new light on previously perplexing patterns in the natural world (why there are no native land mammals on oceanic islands, for example).

Since its inception, over a decade ago, the theory of ID has provided no illuminating insights into biology. The contribution to science of ID is no more useful than the explanation 'it's magic!'

Methodological naturalism in science does not deny the existence of the supernatural. It simply ignores the supernatural because it is outside the realm of what science can explain.

Science can only make testable predictions about things and events that are subject to the laws of nature.

Evolution

ID is a 'God in the Gaps' Argument from Ignorance

An argument from ignorance involves claiming that a lack of evidence for one view is evidence for another view. Thus when proponents of ID point to a perceived flaw in evolutionary theory, or to some feature of the world that evolutionary scientists have not yet explained, they firstly suggest that scientists simply cannot explain this feature in evolutionary terms, and then take this as suppport for their own account. In response to this, evolutionary scientists point out that an absence of evidence for some evolutionary change is not itself positive evidence that this change actually did not occur.

A 'God in the gaps' argument is a particular version of this: it points to a gap in scientific knowledge, this gap is explained as the handiwork of a god, and may then be taken to prove a god's existence. Again, when proponents of ID point to gaps in the evolutionary account, things that have not yet been explained, it is generally suggested that science cannot explain the cell structure, DNA, or whatever is at issue. It is then suggested that, for this reason, the thing has been designed. But in fact, in most of these examples science is starting to explain more and more.

Intelligent Design

ID is Not a 'God in the Gaps' Argument from Ignorance

ID is not a 'God in the gaps' argument from ignorance. An intelligent design inference—the inference that something has been designed—is based on positive evidence. The scientist looks for evidence of specified complexity, as well as for evidence of complex design features—features which have been used elsewhere in different systems, and hence are being re-used by the designer in this one.

And, as with any good scientific theory, a design inference can be falsified, either by showing a lack of design in the structure, or by demonstrating that unguided natural processes can create the pattern in question.

It is true that ID is highly critical of evolutionary theory, but this is due to the historical nature of both enterprises. In historical sciences, theories cannot be tested by predicting outcomes or repeating experiments. Rather, a theory is tested by comparing its explanatory power against competing theories, to determine the best explanation. This is the reason ID points out flaws in evolutionary theory, in order to argue that ID is, in fact, the best explanation.

Intelligent Design

ID is Not a Science Stopper

The discovery that the earth was round stopped certain avenues of enquiry because they were show to be fruitless. At the same time it opened up new avenues of enquiry. Once the scientific community accepts ID, scientists can stop wasting time speculating about where life came from, and focus on the more interesting question of how life works.

Beyond this, does the theory of ID make predictions that open up new areas of enquiry? The idea that parts of nature were purposefully designed does prompt a different attitude towards it.

In a previous chapter we looked at the idea of junk DNA: parts of a gene that have no discernible affect on the appearance and behaviour of the organism. If a gene was purposefully designed, however, this DNA must in fact have a purpose. This is what ID predicts. It is up to scientists, then, to work out what this purpose is.

Evolution

ID is a Science Stopper

If we grant that the universe is the result of intelligent design, what is the next step? For example, assume a particular eco-system is the creation of an intelligent designer. Unless this intelligent designer is human, and unless we have some experience with the creations of this and similar designers, how could we proceed to study this system?

If all we know is that a system is the result of ID but that the designer is of a different sort of being than we are, how would we proceed to study this system? It is presumptuous to assume that an intelligent designer would create an eye the way a human engineer would design a similar system with a similar function. Appealing to an intelligent designer to explain some complex system is to explain nothing about that system's relation to its alleged designer. The theory illuminates nothing.

In contrast to the theory of evolution, the theory of ID, since its inception over a decade ago, has inspired no nontrivial experiments and has provided no illuminating insights into biology. In terms of its contribution to science, the theory of ID is no more useful than the explanation 'it's magic!'

Where to Stand?

In order to personally evaluate the debate between ID and evolutionary theory, it might be possible to separate two bases of assessment. The claims of the debate could be assessed firstly in intellectual/scientific terms, to determine which side marshals more compelling evidence and is better science. The debate could then be assessed, in relation to one's own religious viewpoint.

Unfortunately these two bases can be hard to separate. For example, Alvin Plantinga notes that, if your religious viewpoint is atheism (divine entities do not exist), then evolution is 'the only game in town.' It is the only available answer to the question 'where did life come from?' In this case, your religious viewpoint affects your assessment of scientific validity.

If atheism is indeed a religious viewpoint, it is poles apart from, say, the Flat Earth Society, whose members hold that the Bible describes the earth as flat, hence it is so. Not surprisingly, this leads the Society to reject much of science since the middle ages. And between these two extremes ...

Between these two extremes there are a number of religious viewpoints. You may not have worked out which viewpoint you are most sympathetic towards, but the viewpoint you support will clearly affect your assessment of the debate between ID and evolutionary theory.

Let me outline the available options, as described by Eugenie Scott, director of the National Centre for Science Education, in California.in her article 'The Creation/Evolution Continuum.'

THE SPECTRUM OF RELIGIOUS BELIEF

Young Earth Creationism

While they do not interpret the flat-earth passages of the Bible literally, Young Earth Creationists do accept Genesis literally, and calculate the age of the earth on the basis of historical evidence for Noah's flood. They estimate that the earth is between six and ten thousand years old, which conflicts with the findings of modern physics, chemistry and biology. This view was popularised by Henry Morris in 1961, and was the basis for Creation Science, which was taught in some US schools up until 1987.

Old Earth Creationism

Since the 19th century, scientists have presented data which suggests that the earth is very old and religious thought has attempted to accommodate this idea in various ways.

Gap Creationism is the idea that the earth was created as described in Genesis 1:1, but after the appropriate amount of time it was destroyed and recreated in the six-day creation described in Genesis 1:2.

Another attempt to accommodate science to literal readings of the Bible is Day-Age Creationism. On this view, each 'day' of creation is in fact an age of historical time.

Understood in these metaphorical terms, Genesis and the fossil record are in agreement: plants do indeed appear before animals, with humans then emerging as the final crowning glory. There is, however, no such thing as descent with modification (evolution), according to this approach.

Progressive Creationism

According to Progressive Creationism, God created different kinds of animals sequentially, as the fossil record indicates. These 'kinds' allow for genetic variation and evolution within a kind. An original cat, for example, may be the ancestor of both leopards and housecats. Such microevolution is accepted as being the result of evolutionary mechanisms: mutation, natural selection, genetic drift, and so forth.

What this view rejects is the idea that an animal kind could morph into a quite different one (macroevolution). On this view, God created the 'basic body plans' that appeared in the Cambrian explosion and, from there, microevolution fashioned the species we see today.

In terms of the spectrum of religious views being presented here, it is worth noting that Progressive Creationism retains the idea that humans were assigned a 'special' place in creation.

Intelligent Design

This viewpoint more or less accepts evolutionary theory, except in those cases where a biological system is considered too complex to have arisen through purely natural processes. Such complex systems are held to have been designed. It is not immediately clear which aspects of evolutionary theory are ruled out by this criterion.

Certainly microevolution is accepted, but supporters of ID generally follow Progressive Creationism in denying that one 'kind' of animal can give rise to another. The argument given is that 'major body plans' are too complex to have arisen naturally. For example, ID proponent and biochemist Michael Behe accepts that apes and chimps may have a common ancestor (due to the presence in their genes of common pseudogenes, which are genes that are no longer funcional), but it is unlikely that he would include humans in this lineage.

Evolutionary Creationism

and Theistic Evolution

These views hold that evolution is the means by which a god creates life and the universe. This line of thinking accepts the theory of evolution whole-heartedly, although it still leaves open the possibility that a god may have intervened at critical moments (for example, in the creation of humans). But such intervention is an exception to the natural order, a miracle that simply reorients the orderly workings of nature.

This is the view taught in mainline Protestant seminaries, and is the official position of the Catholic Church. In 1996, Pope John Paul II reiterated that God created, evolution happened and humans may indeed be descended from more primitive forms, but the hand of God was needed for the creation of the human soul.

THE SPECTRUM AND SCIENCE

One way to begin analysing the spectrum of religious viewpoints outlined above is to ask how each view affects the current practice of science. If a view is accepted, how much science can still be practised? Beyond this, which views allow scientists to adhere to the idea of Methodological Naturalism in their work?

Methodological Naturalism is the idea that science should explain the natural world in natural terms, in terms of matter and energy, rather than invoking supernatural entities like a designing god.

Accepting Science

Young Earth Creationism conflicts with the findings of modern physics, chemistry and biology. Accepting this view would mean that scientific study would frequently generate conflicts with one's beliefs, in which case it is hard to see how much science could be done.

Old Earth or Day-Age Creationism, according to which God created the species more or less as they are found in the fossil record, is more compatible with modern science. However, this view still entails a flat rejection of evolution, which is a key concept in biology, genetics and elsewhere. This would make study in these areas difficult.

Progressive Creationism and ID accept microevolution, change within species. This is compatible with studying biology, genetics, and so forth, with disagreement being limited just to the larger evolutionary story about macroevolution.

Finally, Evolutionary Creationism and Theistic Evolution are altogether compatible with modern science, although a scientist who accepted either view might be more likely to proclaim something a miracle than one who did not.

Accepting Methodological Naturalism

Methodological Naturalism involves framing scientific explanations without appealing to the hand of a god. Obviously, the spectrum of religious beliefs is graded precisely in terms of how much of a creative role is assigned to a supernatural entity.

Young and Old Earth Creationism hold that God created the world in a particular time-frame and in a particular sequence, and would allow scientists to frame naturalistic explanations only if these fit into this time-frame and sequence.

Progressive Creationism would presumably accord with ID, which, as we have seen, accepts naturalistic explanations except in the case of complex systems, which are held to have been designed and which include the 'body plans' of animal kinds.

Finally, Evolutionary Creationism and Theistic Evolution altogether accept naturalistic explanations in science. These viewpoints leave open the possibility that some things in the world may not, in fact, be explicable in scientific terms, and indeed the origin of life, the universe and everything may be one such miracle. Nevertheless, the role of science is precisely to seek naturalistic explanations, even if it ultimately may not be able to explain everything. However, if this view is accepted what role, then, is left for a god?

A ROLE FOR GOD?

The role of God in most of the above religious viewpoints is clear. In the case of Evolutionary Creationism and Theistic Evolution, however, the role of God is less clear. Pope John Paul II says God created, evolution happened.

Does this imply that a god set the process of evolution in motion and then simply let it unfold, without further intervention?

Certainly the Catholic Church accepts the existence of miracles, canonising saints on the basis of them, and these miracles are precisely further divine interventions in the world. How much, how often and in what ways does this God intervene?

This issue of divine intervention is at the core of the debate between ID and evolutionary theory. Yet it would seem to be an issue that, ultimately, can only be decided on the basis of one's personal faith and commitment.

I hope this book has provided useful information about ID, evolutionary theory, and the debate between them. But now it is up to you, the reader, to reflect on your own personal beliefs in order to work out where you stand regarding the debate between ID and evolutionary theory.

Bibliography of Works Cited

Behe, Michael, Darwin's Black Box: *The Biochemical Challenge to Evolution* (Simon and Schuster, 1996)

Darwin, Charles, *On the Origin of Species by Means of Natural Selection, or The Preservation of Favoured Races in the Struggle for Life* (1859, Harvard, 1909)

Dembski, William, *The Design Inference: Eliminating Chance Through Small Probabilities* (Cambridge University Press, 1998)

Dembski, William, *No Free Lunch: Why Specified Complexity Cannot Be Purchased Without Intelligence* (Rowman and Littlefield, 2001)

Gould, Stephen, *Hen's Teeth and Horse's Toes* (Norton, 1994)

Harris, William and John Calvert, *'Intelligent Design: The Scientific Alternative to Evolution'* (National Catholic Bioethics Quarterly, Autumn, 2003)

Hoyle, Fred, and Chandra Wickramasinghe, *Evolution from Space* (Dent, 1981)

Johnson, Phillip, *Darwin on Trial* (Regnery, 1991)

Kerkut, Gerald, *Implications of Evolution* (Pergamon Press, 1960)

Miller, Kenneth, *Finding Darwin's God: A Scientist's Search for Common Ground between God and Evolution* (Cliff Street Books, 1999)

Milner, Richard and Vittorio Maestro, *Intelligent Design? A special report from Natural History Magazine* (www.actionbioscience.org/evolution)

Meyer, Stephen, Darwinism, *Design and Public Education* (Michigan State University Press, 2003)

Morris, Henry and John Whitcomb, *The Genesis Flood: The Biblical Record and its Scientific Implications* (P & R, 1961)

Paley, William, *Natural Theology, or Evidences of the Existence and Attributes of the Deity collected from the Appearances of Nature* (1802 Oxford University Press, 2006)

Plantinga, Alvin, *'When Faith and Reason Clash: Evolution and the Bible'* (Christian Scholar's Review XXI:1, September 1991)

Richards, Jay and Guillermo Gonzalez, *The Privileged Planet: How Our Place in the Cosmos is Designed for Discovery* (Regnery, 2004)

Scott, Eugenie, *The Creation/Evolution Continuum* (www.ncseweb.org/resources/articles)

Haxton, Charles, *The Mystery of Life's Origin: Reassessing Current Theories* (Lewis and Stanley, 1984)

Wells, Jonathan, *Icons of Evolution: Science or Myth? Why Much of What We Teach About Evolution Is Wrong* (Regnery, 2000)

Further Reading

Evolution
Websites

Antievolution: The Critic's Resource (www.antievolution.org)

Darwin Online (www.literature.org/authors/darwin-charles/the-origin-

Evolution Website: BBC (www.bbc.co.uk/education/darwin)

Intelligent Design (skepdic.com/intelligentdesign)

Kitzmiller v. Dover Area School District trial documents
(en.wikipedia.org/wiki/Kitzmiller_v._Dover_Area_School_District_trial_docu

National Center for Science Education (www.natcenscied.org)

The Scopes Trial (www.law.umkc.edu/faculty/projects/ftrials/scopes)

Talk Origins (www.talkorigins.org)

Understanding Evolution (evolution.berkeley.edu)

Books and Articles

Forrest, Barbara and Paul Gross, *Creationism's Trojan Horse: the Wedge of Intelligent Design* (Oxford Press, 2004)

Gould, Stephen, 'Genesis vs. Geology' (*Atlantic Monthly*, September 1982)

Pennock, Robert, *Intelligent Design Creationism and Its Critics: Philosophical, Theological and Scientific Perspectives* (MIT, 2001)

Pennock, Robert, *Tower of Babel: The Evidence Against the New Creationism* (MIT, 1999)

Scott, Eugenie, *Evolution vs. Creationism: an Introduction* (Greenwood, 2004)

Young, Matt and Taner Edis, *Why Intelligent Design Fails: a Scientific Critique of the New Creationism* (Rutgers, 2006)

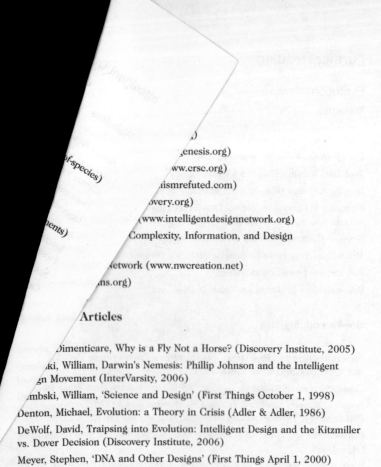

)

enesis.org)

ww.crsc.org)

ismrefuted.com)

overy.org)

(www.intelligentdesignnetwork.org)

Complexity, Information, and Design

Network (www.nwcreation.net)

ns.org)

of-species)

nents)

Articles

Dimenticare, Why is a Fly Not a Horse? (Discovery Institute, 2005)

ki, William, Darwin's Nemesis: Phillip Johnson and the Intelligent gn Movement (InterVarsity, 2006)

mbski, William, 'Science and Design' (First Things October 1, 1998)

Denton, Michael, Evolution: a Theory in Crisis (Adler & Adler, 1986)

DeWolf, David, Traipsing into Evolution: Intelligent Design and the Kitzmiller vs. Dover Decision (Discovery Institute, 2006)

Meyer, Stephen, 'DNA and Other Designs' (First Things April 1, 2000)

Further Reading

Evolution
Websites

Antievolution: The Critic's Resource (www.antievolution.org)

Darwin Online (www.literature.org/authors/darwin-charles/the-origin-of-species)

Evolution Website: BBC (www.bbc.co.uk/education/darwin)

Intelligent Design (skepdic.com/intelligentdesign)

Kitzmiller v. Dover Area School District trial documents
(en.wikipedia.org/wiki/Kitzmiller_v._Dover_Area_School_District_trial_documents)

National Center for Science Education (www.natcenscied.org)

The Scopes Trial (www.law.umkc.edu/faculty/projects/ftrials/scopes)

Talk Origins (www.talkorigins.org)

Understanding Evolution (evolution.berkeley.edu)

Books and Articles

Forrest, Barbara and Paul Gross, *Creationism's Trojan Horse: the Wedge of Intelligent Design* (Oxford Press, 2004)

Gould, Stephen, 'Genesis vs. Geology' (*Atlantic Monthly*, September 1982)

Pennock, Robert, *Intelligent Design Creationism and Its Critics: Philosophical, Theological and Scientific Perspectives* (MIT, 2001)

Pennock, Robert, *Tower of Babel: The Evidence Against the New Creationism* (MIT, 1999)

Scott, Eugenie, *Evolution vs. Creationism: an Introduction* (Greenwood, 2004)

Young, Matt and Taner Edis, *Why Intelligent Design Fails: a Scientific Critique of the New Creationism* (Rutgers, 2006)

Intelligent Design

Websites

Access Research Network (www.arn.org)

Answers in Genesis (www.answersingenesis.org)

Center for Science and Culture (www.crsc.org)

Darwinism Refuted (www.darwinismrefuted.com)

Discovery Institute (www.discovery.org)

Intelligent Design Network (www.intelligentdesignnetwork.org)

International Society for Complexity, Information, and Design (www.iscid.org)

Northwest Creation Network (www.nwcreation.net)

Origins (www.origins.org)

Books and Articles

Darwin, Dimenticare, Why is a Fly Not a Horse? (Discovery Institute, 2005)

Dembski, William, Darwin's Nemesis: Phillip Johnson and the Intelligent Design Movement (InterVarsity, 2006)

Dembski, William, 'Science and Design' (First Things October 1, 1998)

Denton, Michael, Evolution: a Theory in Crisis (Adler & Adler, 1986)

DeWolf, David, Traipsing into Evolution: Intelligent Design and the Kitzmiller vs. Dover Decision (Discovery Institute, 2006)

Meyer, Stephen, 'DNA and Other Designs' (First Things April 1, 2000)